SEARCHING FOR ALIENS

by Jenny Mason

12 STORY LIBRARY
MORE TO EXPLORE

www.12StoryLibrary.com

12-Story Library is an imprint of Bookstaves.

Developed and produced for 12-Story Library by Focus Strategic Communications Inc.

Library of Congress Cataloging-in-Publication Data
Names: Mason, Jenny (Children's author), author.
Title: Searching for aliens / by Jenny Mason.
Description: Mankato, Minnesota : 12-Story Library, [2022] | Series: Aliens! |
Includes bibliographical references and index. | Audience: Ages 10–13 | Audience: Grades 4–6
Identifiers: LCCN 2020014919 (print) | LCCN 2020014920 (ebook) | ISBN 9781632359353 (library binding) |
ISBN 9781632359704 (paperback) | ISBN 9781645821014 (pdf)
Subjects: LCSH: Life on other planets. | Extraterrestrial beings—Juvenile literature.
Classification: LCC QB54 .M378 2022 (print) | LCC QB54 (ebook) | DDC 576.8/39—dc23
LC record available at https://lccn.loc.gov/2020014919
LC ebook record available at https://lccn.loc.gov/2020014920

Photographs ©: sdecoret/Shutterstock.com, cover; CSIRO/CC3.0, 1, 4; Jurik Peter/Shutterstock.com, 5; Lena_graphics/Shutterstock.com, 5; Imaginechina Limited/Alamy, 6; Andrew Z. Colvin/CC3.0, 7; Imaginechina Limited/Alamy, 7; NASA Astrobiology/YouTube.com, 8; NASA/JPL-Caltech, 8; NASA/ESA/G. Bacon (STSci), 9; 3Dstock/Shutterstock.com, 9; Paulo Afonso/Shutterstock.com, 10; M2 Photography/Alamy, 11; Johan Swanepoel/Shutterstock.com, 11; U.S. Department of Energy, 12; Amalex5/CC4.0, 13; MasPix/Alamy, 13; NASA/JPL-Caltech/S. Stolovy (Spitzer Science Center/Caltech), 14; NASA Image Collection/Alamy, 15; NASA/JPL-Caltech, 15; NASA Spitzer/YouTube.com, 15; ESA & MPS for OSIRIS Team MPS/UPD/LAM/IAA/RSSD/INTA/UPM/DASP/IDA/CC3.0 IGO, 16; PD, 17; BusinessImage/Shutterstock.com, 17; Sylke Rohrlach/CC2.0, 18; Achille Beltrame/PD, 19; vrx/Shutterstock.com, 19; Alex Mit/Shutterstock.com, 20; lev radin/Shutterstock.com, 20; Geremia/PD, 21; NASA/Frankie Martin, 21; NASA/JPL-Caltech/Wendy Stenzel, 22; NASA/Goddard Space Flight Center/Pat Izzo, 23; Algol/Shutterstock.com, 23; critterbiz/Shutterstock.com, 24; Rainer Fuhrmann/Shutterstock.com, 25; Daniel Eskridge/Shutterstock.com, 25; Jaminnbenji/Shutterstock.com, 26; Dickenson V. Alley/CC4.0, 27; Triff/Shutterstock.com, 27; Professor Simon Holland/YouTube.com, 28; Hank Leukart/YouTube.com, 28; SPDO/TDP/DRAO/Swinburne Astronomy Productions/CC3.0, 29

About the Cover

Satellite dishes scan the sky.

Access free, up-to-date content on this topic plus a full digital version of this book. Scan the QR code on page 31 or use your school's login at 12StoryLibrary.com.

Table of Contents

Go First: The SETI Institute 4
Eye Spy: China's FAST Achievement 6
In the Red: Speeding Up the Search 8
A Stellar Idea: Allen Telescope Array 10
Getting Sirius: An Alien-Hunting College 12
Total Eclipse: The Latest Tools 14
The 1924 Radio Silence 16
A Closer Look: Technosignatures vs. Biosignatures 18
A Big Project: Breakthrough Listen 20
Game Changer: NASA Hunts Again 22
Look Smart: Spotting Alien Intelligence 24
Listen Up: Nikola Tesla Hears Martians 26
Above and Beyond 28
Glossary 30
Read More 31
Index 32
About the Author 32

1

Go First: The SETI Institute

In 1969, a radio telescope received signals from Apollo 11 during the Moon landing.

Going first often means going alone. Years ago, one group was the first to hunt for aliens in space. That group is the Search for Extraterrestrial Intelligence (SETI) Institute.

When it formed in 1984, the Institute's goal seemed silly. It also seemed impossible. Still, they pointed radio telescopes at stars. They gathered the electrical and magnetic signals flowing through space. Surely, life existed beyond Earth.

In 1995, two Swiss astronomers discovered that other stars had planets. Since then, thousands of these exoplanets have been found.

200

Number of stars the SETI Institute first tracked with Project Phoenix in 1995

- Project Phoenix began after the US Congress blocked NASA from alien searches.
- The Institute borrowed time on busy radio telescopes around the world.
- Radio telescopes can detect radio waves from stars, black holes, and galaxies.

Finding life in the universe is no longer silly. Scientists everywhere are hunting. The SETI Institute now leads the way.

IN THE ZOO

Imagine aliens crowded around Earth. Their technology makes them invisible. They watch humans like animals in a zoo. That is one theory behind why no one has found life except on Earth.

2 Eye Spy: China's FAST Achievement

Aerial view of FAST.

In 2016, China hit the power switch on FAST. The Five-hundred-meter Aperture Spherical Radio Telescope is the largest in the world. FAST's 500-meter dish is over 1,600 feet across. That is about the length of 5 soccer fields.

FAST is known among the Chinese as *Tianyan* or *Eye of Heaven*. Like all radio telescopes, it sees the invisible. It senses energy waves from outer space. Different energies radiate, or beam. These radio signals reveal galaxies. Others reveal exoplanets with magnetic fields. Earth's magnetic field allowed life to evolve.

FAST is the best device to detect Fast Radio Bursts (FRBs). These bursts originate beyond the Milky Way Galaxy. No one knows what causes them—intelligent life? Maybe FAST will help find the answer.

THINK ABOUT IT

What kinds of life might live in the Milky Way Galaxy? Plants? Animals? Fungi? Or something never known before?

The construction of FAST, 2014.

1.2 billion

Chinese yuan ($182 million in US dollars) spent to build FAST

- The bowl-like dish is made of 4,450 panels.
- FAST was constructed in an ancient crater. The site was chosen because it is flood-proof.
- The telescope can sense the edge of the known universe.

3

In the Red: Speeding Up the Search

Jacob Haqq-Misra loves red dwarfs. Not the stumpy fantasy creatures. Red dwarf stars. Planets that orbit red dwarfs are often tidally locked. That means they are like our Moon. One side always faces forward.

The side facing the star gets constant heat and daylight. The other side is always cold nighttime. These planets have a middle zone, or halo. Astronomers think these halos can support life.

Three planets and a red dwarf sun.

Two planets and a red dwarf sun.

8 billion

Age of the TRAPPIST-1 red dwarf star

- TRAPPIST-1 is twice as old as our Sun. It is likely to shine for 12 trillion years.
- Red dwarfs tend to be smaller and cooler than our Sun.
- Their gentle heat and slow burn make them ideal places for life. Good news for the seven planets orbiting TRAPPIST-1.

Haqq-Misra works for the Blue Marble Space Institute of Science. Blue Marble scientists connect online. This global teamwork speeds up the search for intelligent life.

LIFE ON THE MOON

In 2019, an Israeli lunar lander crashed on the Moon. On it was an entire tardigrade colony. Tardigrades are indestructible. Boiling liquids? Freezing cold? The vacuum of space? None of that fazes these microscopic creatures. If they survived, they are true aliens now.

4

A Stellar Idea: Allen Telescope Array

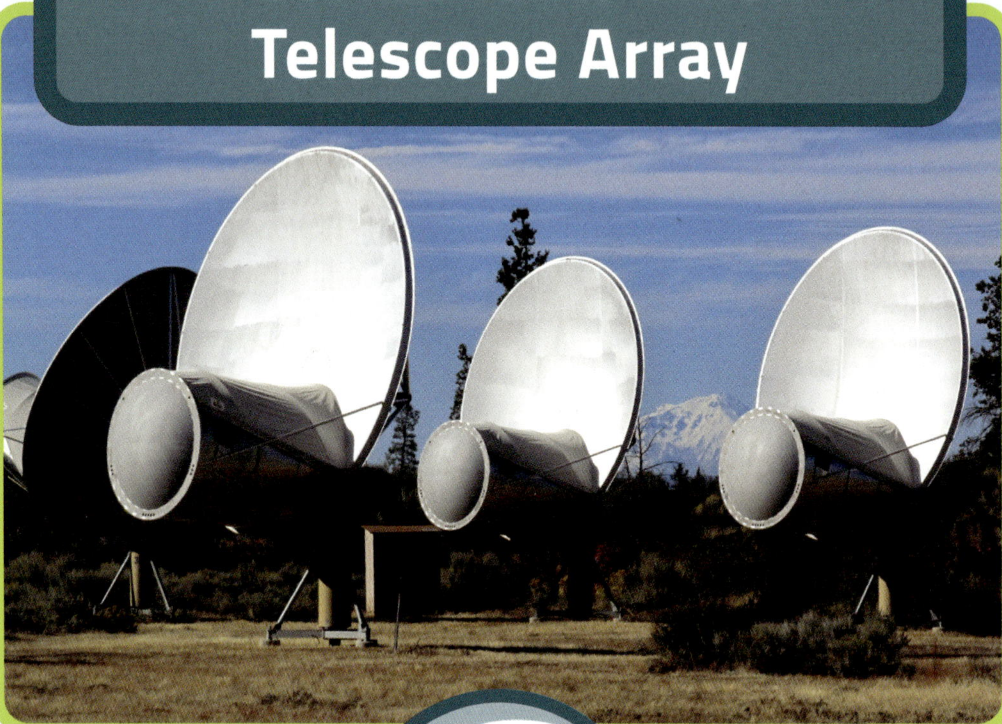

The ATA telescopes are located in Hat Creek, California.

Imagine an eye the size of two football fields. That is what scientists hope to have when they complete the Allen Telescope Array. The ATA telescopes will speed up alien hunting.

Before the ATA, SETI astronomers had to squeeze in time on other telescopes. They had only a few hours or days to search for life in a huge universe. The ATA will monitor space all the time. Once completed, the ATA will boost scientific studies. Instead of 1,000 stars, astronomers will detect over 1 million.

Each ATA telescope is a radio satellite dish. So far, 42 dishes scan outer space

Each ATA dish measures about 20 feet (6 m) across.

for different radio signals. One day, this group of telescopes in California will have 350 dishes. That is a lot of search power.

500,000

Approximate number of junk bits currently orbiting Earth

- The ATA helps other agencies track dangerous space junk.
- Most of it is moving at 17,500 miles per hour (28,000 km/h).
- Radio telescopes can change the energy waves they detect into sounds and pictures.

Getting Sirius: An Alien-Hunting College

Pennsylvania State University is unique. It opened an alien hunting center. It is called PSETI (Penn State Extraterrestrial Intelligence Center). Penn State is the first college to do so.

PSETI reveals changing attitudes. Alien hunting is more popular now. It is also more widely accepted.

The people who search for alien life need lots of help. They have a big riddle to solve. It is called the Fermi Paradox. Enrico Fermi was a brilliant scientist. In the 1940s, he asked other scientists a question. Where is everybody?

Fermi meant everybody in outer space. The universe is almost 14 billion years old. That was enough time to make life on many planets. Fermi

Enrico Fermi won the 1938 Nobel Prize.

Frank Drake, 2017.

wondered, why haven't they found us? Or, why haven't we found them?

1961

Year the Drake Equation became a prediction tool for intelligent life in the universe

- Astronomer Frank Drake made the equation. It estimates how many smart societies might be in space.
- Drake will help PSETI scientists search for super tools.
- Dyson spheres are super tools. They turn stars into mega-batteries.

An illustration of a Dyson sphere system.

6

Total Eclipse: The Latest Tools

The Milky Way Galaxy may have trillions of planets. It's hard to believe we will search each one for aliens. Luckily, NASA is developing several unbelievable options.

Astronomers search for exoplanets near bright stars. To see, space telescopes need a light blocker, or starshade. NASA engineers are designing a giant, mechanical "sunflower." It is as big as a baseball diamond. This starshade will float in front of a telescope and eclipse (block) the star.

The coronagraph is another strange invention. It sits inside a telescope. It is no

An illustration of a starshade.

bigger than a jelly jar lid. The coronagraph has a flexible mirror. The mirror jiggles to remove the pesky starlight.

4

Number of Earth-sized planets that could fit between a telescope and its starshade

- Imagine the shade like your hand. You reach out to block the Sun.
- NASA learned origami to launch the giant shade.
- The sunflower will fold into a 16-foot (5-m) disk. It will bloom in space.

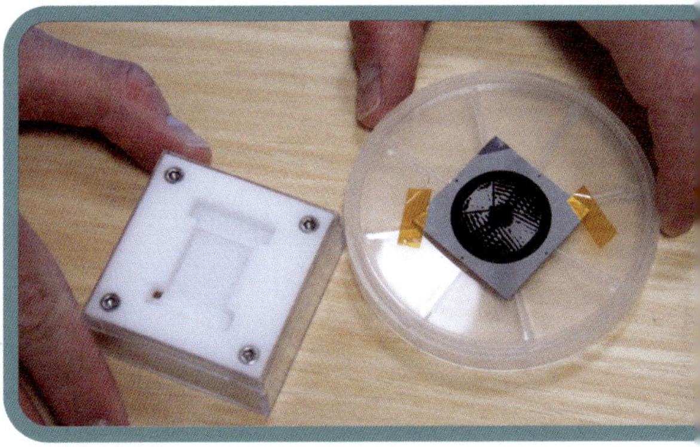

LOOK AHEAD

Lyman Spitzer Jr. invented space telescopes. He also imagined the first starshade in 1962. Jagged edges would direct the light waves away from the telescope's viewfinder.

7

The 1924 Radio Silence

Every hour on the hour. That's how often people across the US were told to shut off their radios. The radio silence spanned from August 21 to August 23, 1924.

Those particular days mattered. The distance between Earth and Mars would be shortest. That was the best time to detect any Martian communication.

Mars.

THINK ABOUT IT

Why is life not likely to exist outside the Goldilocks Zone?

The US Naval Observatory helped. It flew a powerful radio receiver into the sky. William F. Friedman was on hand to listen. He was the chief cryptographer (code breaker) for the US Army. Maybe Martians spoke in codes. Friedman would translate it. The public was excited. Scientists were excited. Still, the project heard nothing. They did not know then that Mars is uninhabited.

William F. Friedman, 1950s.

35 million

Number of miles (56 km) between Mars and Earth when their orbits aligned in 1924

- The two planets swing by each other every 779 days.
- Both planets share the Sun's "Goldilocks Zone." That area is not too hot and not too cold.
- To find alien life, scientists search the Goldilocks Zone around other stars.

Earth and Mars.

8

A Closer Look: Technosignatures vs. Biosignatures

Astronomers hunting for alien life look for two basic clues. *Biosignatures* include anything produced by living creatures. Bodily gas is one example. Enzymes (chemicals used for eating) are another. For example, a sea slug feeds by spitting enzymes onto a sponge to soften it. Then it sucks up the food.

Technosignatures are signs of smart creatures using technology. These include roads or buildings. Bright city lights are another example. Maybe aliens pollute.

In 2015, astronomers thought they found a technosignature. Boyajian's Star (or KIC 8462852)

1877

Year that some astronomers claimed Mars had technosignatures

- Italian astronomer Giovanni Schiaparelli discovered canyons on Mars. He called them *canali*.
- Other scholars insisted only smart beings can build canali.
- In 2015, NASA found organic materials on the dwarf planet Ceres.

flickered in odd ways. Maybe a super technology caused it. For example, a Dyson sphere is an imagined super tool. Perhaps aliens used a sphere to pull energy from Boyajian's Star. Scientists now think ordinary dust creates the weird flicker.

A Dyson sphere.

RECIPE FOR SUCCESS

In 1952, two scientists mixed water, heat, gas vapors, and electric sparks. These ingredients acted the way oceans, sunlight, and lightning affected Earth long ago. Weeks later, the water turned red-brown. It was full of life-building proteins. The recipe for life was surprisingly simple.

9

A Big Project: Breakthrough Listen

The Milky Way, a spiral galaxy.

What's the fastest way to find alien life in a vast universe? Link the world's biggest radio telescopes. Make them search all at one time. This great idea is also very expensive.

Luckily, billionaire Yuri Milner will spend $100 million dollars to do just that. The project is called Breakthrough Listen. It scans one million stars. Breakthrough will listen to 100 galaxies.

Yuri Milner, 2016.

Breakthrough Listen monitors for alien radio signals and lasers. It uses the Green Bank Telescope, the world's largest steerable telescope. The Parkes Telescope and China's FAST also help listen. In 2019, a signal was detected. It came from Proxima Centauri, the closest star to our Sun. Breakthrough doesn't know what made the signal. It could be aliens. Or it could be a new object in space.

The Green Bank Telescope.

10,000

Number of new exoplanets that NASA's TESS (Transiting Exoplanet Survey Satellite) may find

- TESS is a space telescope. It launched on April 18, 2018.
- Breakthrough Listen partnered with TESS a year later.
- Every day, Breakthrough Listen gathers data that the previous system took a year to get.

10 Game Changer: NASA Hunts Again

The Kepler Telescope.

NASA stopped searching for alien life in 1993. The US government cut the budget. It was a huge blow to space science. Luckily, NASA's Kepler Telescope discovered exoplanets circling other stars. *Exo* means outside our solar system.

The news was stunning. In 2018, Congress gave NASA funds to resume searching. NASA launched the James Webb Space Telescope (JWST). It looks for large planets, or mega-Earths.

It takes two trucks to ship the JWST from place to place.

In the 2020s, NASA launched the Roman Space Telescope. This telescope can detect oxygen and water vapor. To do so, it studies the light bouncing off planets. White light splits into a rainbow. Different colors reveal different gases. Oxygen. Methane. Mixed gases in smog (air pollution). These gases are good signs of life.

4,100

Number of confirmed exoplanets found so far

- Astronomers estimate that 300 billion stars swirl in the Milky Way Galaxy.
- Each star may have 100 or more planets and moons in orbit.
- The math suggests 30 trillion exoplanets may exist.

FLYING DRAGONS ON MEGA-EARTH

Mega-Earths have extra gravity, or supergravity. Scientists suppose that creatures on these planets could fly. For example, dragon-like snakes with airbags, or bladders, in their bodies could defy the gravity.

11 Look Smart: Spotting Alien Intelligence

Humankind considers itself very intelligent. We build roads to get around easily. Some exoplanets are big ice balls. To get around, the aliens might develop pipes and tunnels. Is that just as intelligent as using streets?

New studies of animal behaviors help SETI Institute scientists better understand alien smarts. For example, Humpback whales blow bubble nets to trap fish. Chimps form alliances to get to the top of a social group.

Slime molds are smart without brains. They can solve mazes. They memorize. They can even recognize patterns. That means they can guess at the future. People do all

Humpback whales in the waters of Alaska feed using bubble nets.

Slime mold.

this, too. Sharpening their intelligence-finding skills on Earth will help scientists find intelligence in space.

THINK ABOUT IT

Many intelligent creatures use camouflage, or disguises. How could we spot alien life if it disguised itself?

500 million

Number of neurons in an octopus brain.

- Half of those neurons are located in the octopus's tentacles. Neurons send and receive information.
- Octopuses can invent and plan. They also like to decorate their homes.
- Scientists think the octopus brain may be the most like alien brains.

12

Listen Up: Nikola Tesla Hears Martians

Pikes Peak.

In 1899, Nikola Tesla detected outer space signals. The famous inventor was on top of Pikes Peak. That Colorado mountain soars over 14,000 feet (4,300 m) high. It is a snowy, stormy place. It was perfect for Tesla's experiments.

Tesla had already mastered moving electricity through wires. Now he wanted wireless technology. Just beam electricity through the air. He was working alone in his laboratory. An instrument picked up the odd signals.

Nikola Tesla in his laboratory with his high-voltage generator, 1899.

Tesla guessed the signals came from Mars. Maybe smart beings there were communicating. He believed it would be possible to send a message to Mars. Tesla shared the discovery with the public in 1901. It was exciting news. At the time, most people believed in life on Mars. Years later, scientists traced the signals to gas clouds in space.

45

Number of space missions directed at Mars since 1960

- Missions have included robotic probes, rovers, and orbiting satellites.
- Besides the US, many countries aim for Mars. Russia, China, and India are on the list.
- So far, no mission has found evidence that life ever existed on Mars.

A Mars rover.

Above and Beyond

Professor Simon Holland explains Project Argus, 2019.

Join the All-Stars

Anyone can hunt for aliens. First, join Project Argus. It is a team of professional and at-home astronomers. Argus has about 1,500 members across 62 countries. It is part of the SETI Institute.

Next Gen Explorers

The Ad Astra Acadamy partners with low-income schools. Recently, it worked with students in Brazil and Bangladesh. They snapped pictures of the Mars surface. They worked directly with NASA.

Students in Brazil explore Mars using a NASA spacecraft.

Array of Hope

Telescope arrays link many radio telescopes. New Mexico's Very Large Array has 27 telescopes. That is like a single dish the size of New York City. South Africa and Australia are building the Square Kilometer Array (SKA). It will survey space with thousands of dishes.

Glossary

camouflage
The colors or shapes that protect animals from attacks by making them hard to see.

equation
A mathematical statement.

exoplanet
A planet outside Earth's solar system.

neuron
A cell that carries messages between the brain and the body.

origami
The Japanese art of folding paper into animal shapes.

organic
Anything related to material that comes from living things.

paradox
A statement that seems to express two opposite, yet true, ideas.

protein
An organic compound that builds bodily tissues such as skin or hair.

radiate
To extend out in the form of rays.

tardigrade
An almost microscopic animal with eight legs.

telescope
A device you look through to see things from far away.

Read More

Heilman, Christine, and Margaret Parrish, eds. *Space!: The Universe as You've Never Seen It Before.* New York, NY: DK Publishing, 2015.

Simon, Seymour. *Exoplanets.* New York, NY: HarperCollins, 2018.

Tyson, Neil deGrasse. *Astrophysics for Young People in a Hurry.* New York, NY: Norton Young Readers, 2019.

Index

Ad Astra, 28
Allen Telescope Array, 10–11

biosignatures, 18
Boyajian's Star, 18–19
Breakthrough Listen, 20–21

Drake Equation, 13

exoplanets, 21, 22–23

FAST (Five-hundred-meter Aperture Spherical Radio Telescope), 6–7
Fast Radio Bursts (FRBs), 7
Fermi Paradox, 12–13
Friedman, William F., 17

Haqq-Misra, Jacob, 8–9

intelligence, 24–25

Mars, 26–27
mega-Earths, 23
Milner, Yuri, 20

octopuses, 25

Project Argus, 28
Project Phoenix, 5
PSETI (Penn State Extraterrestrial Intelligence Center), 12–13

radio silence, 16–17
red dwarf stars, 8–9

Search for Extraterrestrial Intelligence (SETI) Institute, 4–5
space junk, 11
Spitzer, Lyman Jr., 15
Square Kilometer Array (SKA), 29
starshades, 14–15

tardigrades, 9
technosignatures, 18
telescopes, 14–15, 20–21, 22, 29
Tesla, Nikola, 26–27

Very Large Array, 29

whales, 24–25

About the Author

Jenny Mason has always lived in UFO country. She grew up a few hours from Roswell, New Mexico. She currently lives in Colorado near the UFO Highway. She will go anywhere in search of amazing stories.